C000000748

SCOTCH WHISKY

TED BRUNING

Published in Great Britain in 2015 by Shire Publications Ltd,
PO Box 883, Oxford, OX1 9PL, UK.

PO Box 3985, New York, NY 10185-3985, USA.

E-mail: shire@shirebooks.co.uk   www.shirebooks.co.uk

© 2015 Ted Bruning.

A CIP catalogue record for this book is available from the British Library.

Shire General no. 10.

United Kingdom ISBN-13: 978 0 74781 464 1
United States ISBN: 978 1 78442 057 4
PDF e-book ISBN: 978 1 78442 066 6
ePub ISBN: 978 1 78442 065 9

Ted Bruning has asserted his right under the Copyright, Designs and
Patents Act, 1988, to be identified as the author of this book.

Designed by Alysa Thomas and typeset in Bembo and Helvetica Neue.

Printed in China through Worldprint Ltd.

15 16 17 18 19   10 9 8 7 6 5 4 3 2 1

Acknowledgements:
Alamy, pages 4, 13, 38, 44, 46–7,
50, 86; Abhainn Dearg Distillers,
page 95; Albino Vieira Filhos,
page 26; Leslie Barrie, page 74;
Bridgeman Art Library, page
22; Gordon James Brown, page
108; James Bullock, pages 21,
77; Chivas Bros, page 52; John
Dewar & Sons, pages 49, 51, 59,
63, 64; Anne Burgess, page 45;
Diageo, pages 65, 88, 89; Dick
& Debbie's Travels, page 79; The
Famous Grouse, pages 43, 90, 104;
Glenfarclas, pages 7, 9, 10, 12, 15
(top), 16, 54–5, 56; Government
of Iraq, page 24; Isle of Arran
Distillers, page 94 (bottom);
Kilbeggan Distillers, page 20
(right); Kilchoman Distillers, page
94 (top); US Library of Congress,
page 81; Loch Ewe Distillers,
page 33; Orange County Archive,
page 78; Curt Robinson, pages
6, 8, 15 (bottom); Scotch Whisky
Association, pages 18, 97; Scottish
National Portrait Gallery/
Bridgeman Art Library, page 42;
Shutterstock, cover image and
pages 35, 36, 40–1, 60–1, 68–9,
96, 98 (top and bottom), 100,
101, 102; Wikimedia Commons,
pages 25, 27, 28, 82; William
Grant & Sons, pages 20 (far
right), 66, 67, 70–1, 91, 92, 93.

CONTENTS

HOW SCOTCH
IS MADE

SCOTCH WHISKY is the world's best-selling spirit and Britain's greatest single contribution to global gastronomy. And to its armies of fans in Britain, in Europe, in North America, and now in China, India and Brazil as well, it's much more than just a dram. It's the very spirit of Highland glen and mountain burn, of loch and of peat bog.

For once, the hyperbole is justified: the romance of Scotch is its reality. Just as much as any great French wine, a malt whisky is the ultimate expression of its *terroir*. The peat the malt is dried over, the purity of the spring water, the shape and size of the stills, the oak barrels — a thousand variables make a single malt whisky what it is.

The unromantic definition of Scotch is that it's the distillate of a fermented wash of malted barley to which whole grains of other cereals may be added, distilled to less than 94.8 per cent alcohol by volume (ABV) and aged for at least three years in oak casks of 700 litres or more. The only permitted additives are water and caramel; and the end product may not be sold at less than 40 per cent ABV.

Opposite:
A peaty burn in Aberdeenshire. Phenolic compounds leached into streams can profoundly affect the flavour of the whisky.

Opposite: Barley
ripening at
Glenfarclas in the
gently rolling hills
of Banffshire in the
Speyside region.

Freshly cut peat.

Finally, the whole process up to blending must be carried
out in Scotland.

This summary of the 1988 Scotch Whisky Act is only the
screw-cap on a whole bottle of history. In effect it reconciles
the differences between the very different spirits of which
Scotch is composed: malt whisky and grain whisky. For
it's not the Scotch Whisky Association's five regions of origin

– Highlands and Islands, Lowlands, Campbeltown, Islay and Speyside – that characterise Scotch: the whiskies of each region have nothing in common, and the regions themselves are purely historic. Instead, the two great definers are, first, malt or grain and, second, in the case of malt, the microgeography of each individual distillery. You might be able to identify two similar wines as clarets; but unless you already knew, you'd never guess that Laphroaig and Bunnahabhain both came from Islay. The differences come down to ingredients and processes, and each distillery has its own.

Opposite:
Traditional floor maltings at Laphroaig. Here the moistened barley is heaped up and allowed to start germinating; then gentle fires are lit underneath and it is turned by hand during the drying process.

Checking the grist at Glenfarclas. This is malted barley ground into a coarse flour and ready to be turned into 'wash', or unhopped beer.

Let's begin with malt whisky. It starts life as straightforward unhopped beer, which is mostly water, so the distillery's water supply is an important ingredient in the construction of its whiskies. It might be hard and laden with minerals; it might be soft and pure; it might have flowed through peat bogs and picked up traces of flavour there.

But it's the barley, and in particular the way it's malted, that dominates. A grain of barley is a self-contained package that carries enough energy in the form of starch to get the growing process under way. The package includes an enzyme, diastase, which, once moistened, will turn the starch to sugar to feed the plant while its roots develop. The maltster tricks the barley into germinating by soaking it, and then arrests the process by drying it. The fuel used in the drying is critical: industrial maltsters blow neutral warm air heated by gas or oil into great revolving drums of damp grain, but in the past the choice was straw or furze, charcoal, coke or in some lucky localities peat, burning directly under a bed of germinating grain. All these options produce different flavours, but peat is the most distinctive: a heavily peated malt such as Laphroaig is either deliciously smoky or pungently medicinal, depending on the taste of the drinker.

The malt is then crushed and steeped, exactly as in a brewery, to extract its sugars; but instead of being boiled with hops, the sweet malt syrup goes directly to fermentation

Opposite:
Filling the mash tun at Glenfarclas. In the mash tun the grist is soaked in hot water until it has surrendered all its fermentable sugars. The sweet liquid that results is called 'wort'.

Opposite:
Fermentation in progress in the washback at Glenfarclas. The washback is where yeast is added to the fresh wort to produce a malt liquor of about 8 per cent ABV.

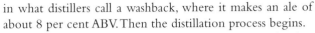

Interior of a still at Glenfarclas. The wash is passed through the still twice. The first pass separates out the worst of the impurities; the second concentrates and refines the spirit to more than 90 per cent ABV.

in what distillers call a washback, where it makes an ale of about 8 per cent ABV. Then the distillation process begins.

A pot still is simplicity itself. It's a bulbous vessel either made of or lined with copper (which eliminates sulphur-based compounds), with a fire underneath and an inverted funnel on top. The wash in the still is gently heated, and the vapour that will eventually be Scotch starts to rise into the funnel or head. This leads to a condenser, traditionally a coil of copper tubing immersed in cold water, via a pipe called a lyne arm. Every factor is a variable: the size and shape of the still, the height and bore of the head, the length of the lyne arm and even the angle at which it's set – all of them affect the separation of the different oils and heavy alcohols (some of which, like methanol and fusel oil, are extremely undesirable) from the ethanol. The general principles are well understood; the exact details aren't. That's why, when stills or their components have to be replaced, the new parts must exactly match the old.

Malt whisky is distilled twice or, in some cases, three times. The first distillation is really a cleaning process during which undesirable compounds are separated out. The liquid that results is called the 'low

Previous page:
Stillheads at
Glenfiddich,
in Moray.

wines'. Once recondensed, the low wines flow through the spirit safe, a locked glass box that allows the stillman to observe the run-off and discard the first and last runnings without actually being able to touch the spirit. To prevent pilfering, the entire system must be closed until duty is paid. The acceptable portion, which is generally around 25 per cent ABV, then goes to the second still for the process to be repeated. Again the first and last runnings, or foreshots and feints, are discarded (and recycled to the first still), while the middle cut is filled into oak casks. The unmatured 'new make' is colourless and fairly neutral: it's the maturation that will give it character.

Here again it's not just the length of maturation but the nature of the barrels that counts. The oak confers tannin, which colours the spirit, and vanillin, which smooths it. Some of the ethanol – the 'angels' share' – will evaporate through the porous wood. Any residual fusel oils, aldehydes and congeners will be broken down over time. The new make will also absorb flavour from the barrel's previous occupant. American distillers use their barrels only once, and many of these make their way to Scotland; the insides of Bourbon casks are charred before filling and therefore have a particular character. Sherry casks are frequently used to 'finish' maturing stocks – that is, they are filled with mature whisky for the last few months before bottling.

That, more or less, is how malt whisky is made. But this accounts for less than half of the whisky we drink.

Spirit safe at Glenfarclas. Once the condensed vapour starts running, the stillman has to decide where to start and stop the 'middle cut'. To prevent the removal and sale of spirit before duty is paid, he has to watch the process through a securely locked window.

Spirit safe at Bruichladdich, Islay.

Grain whisky, of which blended whisky is mainly composed, undergoes a very different process.

The patent (or 'continuous' or 'Coffey') still in which it is made is an ingenious piece of engineering that recycles its own heat. It comprises two columns, the analyser and the rectifier. A continuous stream of wash fermented from cooked maize and other unmalted grains, with only as much barley malt as is necessary to supply the diastase to convert all the starches, is piped into the top of the analyser while steam enters at the bottom. The steam evaporates the wash, and the whole cloud rises to a funnel through which it is piped into the bottom of the rectifier. The stream of wash on its way *to* the analyser first passes through a coil in the rectifier where it encounters the hot mixed vapour on its way back *from* the analyser. The rectifier is in effect a graduated heat exchanger, with ambient wash entering at the top and boiling vapour at the bottom. As the vapour rises and the wash descends, they exchange heat, so the temperature in the rectifying column progressively falls off towards the top. Undesirable vapours condense at different temperatures and are collected as they do so, while the ethanol continues its rise to the top of the column, to be condensed and collected at 90 per cent or more pure.

Can this be whisky? In 1905 the London Borough of Islington decided not. It summonsed a publican for selling blended Scotch, arguing that the product of the continuous still was so different from that of the pot still as not to merit

Opposite:
Casks at Glenfarclas. Maturation is critical: it infuses the spirit with tannin and vanillin from the oak; it allows the breakdown of aldehydes and other unwanted chemicals; and it permits the evaporation of a small quantity of alcohol – the 'angels' share'.

Casks at Deanston,
near Stirling.

Right:
Early Coffey
still preserved
at Kilbeggan
Distillery, West
Meath, Ireland.
This ingenious
heat exchanger,
invented two
hundred years
ago, allows
the continuous
production of
spirit from a mash
containing as little
as 20 per cent
malted barley.

Far Right:
Grain whisky is
hard to obtain but
the taste makes the
effort worthwhile.

Opposite:
Modern column
still at Adnams,
Southwold, Suffolk
– very similar in
appearance to its
ancestor.

the name whisky. The publican was convicted. The industry lobbied for – and in 1908 got – a Royal Commission that eventually declared that if it was made in Scotland, and as long as the grist, of whatever grain, was saccharified by the diastase of malt, that would do. And certainly, at a blind tasting, most people would have difficulty in identifying a well-matured single grain such as Cameron Brig alongside a range of light single malts and premium blends.

SCOTCH ANCIENT
AND MODERN

THE ORIGINS of distilling are much argued over. The science first entered Western literature in the fourth century BC with a brief mention in Aristotle's *Meteorology* as a way of desalinating seawater. At the same time the metallurgists of southern India were isolating zinc by distillation, while the Chinese were distilling medicines and cosmetics from botanicals. None of these societies existed in isolation, but how or even whether knowledge of distillation was transmitted between them is unknown.

There is no doubt, though, that by the late eighth century AD the scientists of the Islamic Enlightenment had realised that liquids of different densities have different boiling points and can be separated by careful heating. The earliest known writer on the subject, Jabir Ibn Hayyan (died 815), was actually experimenting with wine, which, as he was a devout Muslim, was otherwise of no value to him.

The principal use of *araq al-nabidh* (the sweat of wine from the droplets running down the condenser) was in the manufacture of inks, lacquers, medicines and cosmetics

Opposite:
Whisky still at Lochgilphead, painted in 1819 by Sir David Wilkie.

— especially kohl, which gives us our word alcohol. But it was also enjoyed as a drink by many. The Muslim empire was peopled by millions of vanquished Christians, who ran a vast wine industry. The Muslims, being teetotal but tolerant, let them get on with it; and some of them weren't above straying

Abu Nawas, the poet who first described the joys of drinking brandy. This statue is in Baghdad, a great centre of medical and scientific learning during the Islamic Enlightenment.

into Christian-run taverns, in one of which the ninth-century poet Abu Nawas drank a wine 'as hot inside the ribs as a burning firebrand' – arak, or, as it was sometimes called, *ma'ul-hayat*: the water of life. But it was as a medicine rather than a beverage that this water of life, aqua vitae, came to Christendom.

Romantics like to make a mystery of how this happened, but actually it was very straightforward. Salerno, on the cusp of the Christian and Muslim worlds, had – and still has – the oldest medical school in Europe. By the eleventh century it had amassed a vast collection of Latin, Greek, Hebrew and Arabic texts. Its teachers included the foremost physicians of the age, many of whom, such as Constantine the African, were from Muslim-controlled North Africa; its

Constantine the African, a Christian doctor and writer from Carthage, studied in Baghdad and came to the medical school in Salerno in 1077, bringing with him the medical knowledge of the Islamic Enlightenment.

The simplest possible pot still, comprising boiler, stillhead, worm and wormtub. Stills like this are still manufactured by artisanal coppersmiths throughout Southern Europe – this example comes from Albino Vieira Filhos of Aveiro, Portugal.

undergraduates, most of whom were monks, came from all over Christendom. Part of the curriculum from at least 1100, when a text was published on the subject, was the role of spirits in medicine; and, sadly for the Scots and Irish, this knowledge seems to have come to England first. As early as 1130 Adelard of Bath added several encrypted recipes involving aqua vitae to his copy of the tenth-century recipe book *Mappae Clavicula*. There is a hint in Giraldus Cambrensis's *Topography of Ireland* that Irish monks were distilling in the 1170s: an Anglo-Norman knight, Hugh Tyrell, looted a 'great cauldron' from a convent in Armagh and, back at his headquarters at Louth, it set fire to his lodgings and half the town. Could this have been a still? Perhaps. Less equivocal is the routine recording in the Red Book of Ossory – the late fourteenth-century account-book of Bishop Richard Ledred – of the distillation of wine, presumably as a base for medicinal tinctures.

Then we come to one of the most oft-quoted bills of sale in the history of the liquor industry, which is an entry in the Scottish Exchequer Roll for 1494: 'Eight bolls of malt to Friar John Cor wherewith to make aqua vitae'. This record is generally taken to mark the birth of the Scotch whisky industry, in which it is a little premature, for the next record – the grant in 1505 of a monopoly over

Examining a
patient's urine at
the medical school
in Salerno.

distilling in Edinburgh to the Guild of Barber Surgeons –
still places spirits firmly in the pharmacy. But these two
records are nevertheless hugely significant for two reasons.
First, the sale to Friar John is the first mention of the
distillation of ale rather than wine in Britain. Within a few
decades this innovation was to create the whole family of
northern European white spirits: gin, schnapps, vodka and,
in its earliest period, whisky itself. Second, the 1505 grant
shows that the practice of distillation had jumped the
monastery wall; and once the skills and equipment were out
in the secular world they could be operated by anybody
who brewed ale.

But what can it have been like, this proto-whisky? Well,
pretty much indistinguishable from other malt distillates of
the time – unaged, and therefore clear; double-distilled, or it
would not have been potable; and suffused with medicinal
herbs and spices. As late as 1755 Dr Johnson, in his dictionary,
defined usquebaugh as being 'drawn over aromaticks'. He
could easily have been describing gin.

It would also have been far too strong to dose anyone with
unless it was watered down; and perhaps it was the watering
down that persuaded people that this was not only a medicine,
but a beverage. Its medicinal efficacy was nevertheless
not forgotten or ignored: in the much-quoted words of
Ralph Holinshead's *Chronicle* of 1577, 'it cutteth fleume, it
lighteneth the mind, it quickeneth the spirits, it cureth the

Opposite:
James IV of
Scotland, during
whose reign
(1488–1513) the
distillation of aqua
vitae from malt
liquor was first
recorded and the
barber-surgeons
of Edinburgh
were granted a
monopoly over
distillation in
the city.

hydropsie, it pounceth the stone ...' and so on. Particularly useful was that 'it kepyth ... the stomach from wamblying'.

Not only was aqua vitae, to Holinshead, 'a sovereign liquor' against sickness; it was also easy to make. The household washing copper, with the addition of a tight-fitting hood and a coil of copper tubing (the only expensive item), was all the equipment required. And small-scale distilling was a boon to farmers: it meant that in good years they no longer had to sell surplus barley at glut prices but could process and store it almost indefinitely against the lean years.

It was the storage that transformed raw spirit into whisky. In oak barrels the clear spirit took on colour and flavour from the wood, and mellowed through evaporation. And the longer it was kept, the better it became. By the mid-sixteenth century distilling was essential to the farmer's household economy wherever barley was grown. So widespread was it that it had to be temporarily banned following a failed harvest and consequent grain shortage in 1579 (although naturally the ban did not extend to the gentry).

In 1642 civil war broke out in England, and in 1643 both rebel and Royalist parliaments introduced excise duty, as did the Scots in 1644. Duty was ostensibly a temporary measure to pay for the war; but as with income tax 150 years later, it proved too bountiful to be repealed. In England and Wales an efficient collection service rapidly evolved; it's unlikely that, in the very different conditions that prevailed in Scotland, the authorities there were equally well served. There were duty-paying distillers: in fact, the very first distillery to be recorded by name – Ferintosh near Dingwall – was one such. It was burnt down in the 1689–90 Jacobite rising in support of the exiled James II, and as compensation its owner was exempted from duty, which implies that he was paying it in the first place. But he seems to have been an exception.

One of the first actions of the merged government after the 1707 Acts of Union was to make the collection of duties as efficient in Scotland as it was south of the border. This was hardly controversial, but what happened next was. In 1713 the government proposed to extend the malt tax to Scotland, which was expressly prohibited by an article of the Treaty of Union. The Earl of Findlater and Seafield introduced a bill to dissolve the union; and with support from the Whig opposition and anti-union Tories, he lost by just four votes.

The next 110 years were to be a prolonged series of clashes between competing interests. Ranged against and occasionally alongside each other were Westminster, which just wanted money; the English landed gentry, who were making fortunes growing barley for the London gin distillers and wanted to protect their profits; the landed gentry of the Scottish Lowlands, who wanted access to the English gin trade; the landed gentry of the Scottish Highlands, who wanted to protect their legal distilleries against small-scale competition; and finally the Scottish tenant farmers, especially those in the

Highlands, who just wanted the same income from distilling as their sires. Last and definitely least, these may have been in political influence; but they were not in as awkward a position as their landlords, who had mutually exclusive goals. On the one hand, they wanted their own distilleries to flourish, and would have loved to suppress their tenants' untaxed enterprises; on the other, the same tenants depended on the income from their stills to pay their rent.

To complicate matters further, there was always Jacobitism, the continuing support in Scotland for the disinherited descendants of James II. With well-armed clans always ready to rebel, it was no time for government blunders; and in 1725 a blunder duly came. Because the malt tax cost more to collect than it raised, the government proposed to increase it by 3d a bushel. Although the new rate would still have been only half that charged in England, it provoked riots in all the major Scottish towns. In Glasgow the rioters burned down the MP's house and drove out the garrison. It took four hundred dragoons to retake the city; eleven Glaswegians were shot dead.

It should be noted that the rioters were not distillers but consumers. For the farm distillers were not racketeers like Prohibition bootleggers, nor armed gangs like the smugglers of the English coastal districts. Later on, illicit distillation (called bothie distilling after the bivouacs in which it was conducted) became a full-time occupation for many; but at this stage the distillers were still for the most part ordinary tenant farmers.

Opposite:
The Loch Ewe distillery at Aultbea in Wester Ross has been constructed to evoke the illicit 'bothie stills' of the eighteenth century.

All they wanted was to carry on converting the surpluses of good years into a commodity that would see them through the bad, in which they had the support of their communities and even, though perhaps not unequivocally, their landlords. Most of them, given the times, were surely God-fearing and devout; at least one of them, Magnus Eunson of Orkney, was a cleric. They didn't see distilling as criminal; and they wouldn't pay tax if they could avoid it.

In the memoirs of excise officers – called 'gaugers' because of the dipsticks they carried – there is an undercurrent of tension and the threat, sometimes the fact, of violence. But in the folklore of the times the relationship between gauger and smuggler (as illicit distillers were often described, since transporting the 'peatreek' was as big a part of their job as making it) is portrayed more romantically as the loveable rogue versus the easily fooled and ineffectual official. Wily smugglers are always outsmarting the gaugers by concealing kegs in funeral cortèges, or drilling a hole through a ceiling into a confiscated whisky keg that the gaugers are guarding in an upstairs room; and afterwards the rueful gaugers always congratulate the smugglers on their coup. If there really was anything of this sporting spirit in the struggle between lawbreaker and law officer it can only have been because, by the standards of the times, there was very little at stake. The penalty for being caught – confiscation and a fine, when in London a child could be hanged for

Opposite:
The wild country around Loch Ewe, like much of the Highlands, was almost impossible for excisemen or 'gaugers' to police effectively.

pickpocketing – was not hazard enough to trigger the desperate vendettas conducted between revenue officers and smuggling gangs in Kent and Sussex. And in any case, no matter what loss or inconvenience the gaugers might cause, there were never anywhere near enough of them to police the vast and rugged land under their supervision and seriously disrupt the traffic of whisky from glen to township. 'There are many thousands', reported the Commissioner for Excise in 1758, 'who openly transgress.'

For it was the towns that the smugglers principally served, and the affluent bourgeoisie were their best customers – which, in an age of privilege and entitlement, added another layer of complication to the problem of policing. Even at the time it was widely acknowledged that the Highland product was vastly superior to the Lowland, and for a very good reason. Although the Lowlands did have its bothies, the region was dominated by great landed proprietors who wanted only to join their English counterparts in the gin-distilling gold-rush, and sent great quantities of raw spirit south to be rectified. To match the efficiency of the vast English malt distilleries, they built big shallow stills that could process huge volumes of wash very quickly; the wash was as alcoholic as they could make it; and to avoid the malt tax, they mashed in as much unmalted cereal as possible. Highland whisky, by contrast, was meant for drinking, not for rectifying, and was made in much smaller stills from a weaker wash of pure barley malt. It also

Opposite:
Oban was one of the planned towns built by Highland landowners in the eighteenth century to increase the value of their estates. A distillery was very often included in the plans.

generally spent time in oak in the purchaser's cellar. The better-off drank it openly without fear of arrest; and in the early nineteenth century some of it even started finding its way to London, where the Prince of Wales was a particular admirer.

This state of apparent equilibrium, however, could not last; and it was the success of the Lowland distillers in the London market, where in 1785 they sold 800,000 gallons of raw spirit, that broke the peace. The English distillers, who had considerable political clout, demanded protection, while the jealous Highland lairds wanted a level playing field. The latter were very enlightened landlords in the late eighteenth century, founding new towns to improve the condition of their tenants (and to boost the value of their estates). Oban, Grantown-on-Spey, Keith, Fochabers, Tobermory and Fort William were among dozens of settlements planned and constructed in the late eighteenth and early nineteenth centuries; many of them came equipped with legal distilleries. These the developers generally rented to foolhardy entrepreneurs, many of whom were so undercut by the bothie distillers that they soon went out of business. Some of them promptly went back into business – as bothie distillers. The Highland lairds did what they could to protect those of their tenants who operated legally, which included exerting political pressure, and in 1784 they secured the passage of the Wash Act. This set the minimum legal capacity for stills in the Highlands at 20 gallons, with a licence fee of £1 per gallon capacity per year. It also drew an

Opposite:
The distillery at Oban, intended to provide respectable employment for the people of the new town.

artificial line to separate the districts where the concession was in force from the Lowlands, where the much larger stills were licensed at £1 10s per gallon capacity.

Incensed, the Lowlanders fought back and in 1785 got the minimum size of still increased to 40 gallons and, crucially, persuaded the government to ban the sale of Highland whisky outside the concession zone. This did nothing to stop illicit distillation and smuggling, which went on as before until war with France broke out and the government's sudden and urgent need for money brought all negotiations and compromises to an end. In 1797 the licence fee in the Highland region was increased to £6 10s per gallon capacity, which, even though it

was still much less than that paid in the Lowlands, was high enough to ensure the continuation of smuggling and the ruin of even more of the duty-paying Highland distillers.

But conditions in the Highlands were changing in the favour of the lairds. Since the brutal quelling of Bonnie Prince Charlie's uprising in 1745, the region had been experiencing gradual depopulation as small (and potentially rebellious) farmers were evicted from their arable patches to make way for livestock. The great wave of forced clearances, when crofts were simply burnt down and populations were exported wholesale to Canada, came between 1790 and 1820; but by then the process of resettling country people in new

Tobermory was another of the planned towns built by Highland lairds.

towns such as those mentioned above was already well under way. Gradually the supply of surplus barley dwindled, exacerbated by a series of crop failures in the last years of the eighteenth century; the market – or at any rate the local market – dwindled commensurately; and the social pattern that ensured a smuggler could either trust or intimidate his neighbours was disrupted by urbanisation. Secret distilling in towns and cities continued – Edinburgh was reckoned to possess four hundred illegal stills in 1777 – but with potential informers only the thickness of a party wall away, it was a much riskier business. At the same time, following a Royal Commission of 1797–9 to which Lowland distillers complained loudly, with perhaps a touch of exaggeration, of being severely undercut by Highland smugglers, the gaugers' numbers and powers were increased, and Highland magistrates were directed to show less leniency.

Finally, though, it was a Highland landowner – and one of the greatest of them at that, the Duke of Gordon – who set off the train of events that fatally wounded the tradition of smuggling. The Highland lairds were stung into action by the successful lobbying of the English distillers, who in 1814 persuaded the government to enact new regulations that, said a contemporary, amounted to 'a complete interdict' on distilling in the Highlands. As well as imposing a substantial increase in the licence fee, the new regulations outlawed stills of below 2,000 gallons capacity in the Lowlands and 500 gallons in the

Opposite:
The most famous 'gauger' was undoubtedly Rabbie Burns, who used the influence of his patrons to get a job as an excise officer at Dumfries in 1789. He remained in the post, at an annual salary of £50, until his death in 1796.

Glenturret, near Crieff in Perthshire, claims to be Scotland's oldest distillery. It was licensed in 1818, but illicit distilling had been taking place since at least 1775. It is today the 'brand home' of The Famous Grouse.

Highlands and also increased the minimum permissible strength of the wash. The result would have concentrated distilling in the Highlands in the hands of the few proprietors who could source enough malt and fuel to fill and fire such

The Emigrants by sculptor Gerald Laing at Helmsdale, Sutherland, commemorates the thousands of small tenant farmers expelled from their land during the Clearances of the first half of the nineteenth century.

Glen Garioch, established in 1797 and another of the remaining pre-Excise Act distilleries.

prodigious stills; and even then they could have produced only spirit for rectifying. Two years later the Highland lairds succeeded in procuring the Small Stills Act, which effectively reversed the 1814 regime. Then in 1820 Gordon made a stirring speech in the Lords in which he promised that the lairds would suppress illicit distilling if legal ventures were allowed to trade profitably. Yet another commission of enquiry – the fifth – spent three years deliberating; and in 1823 the Excise Act was passed, sharply reducing both the malt tax and excise duty; confirming the smallest still capacity at 40 gallons; and, crucially, allowing the Highland distillers to 'export' their product outside the concession zone.

Strathisla, Keith, licensed since 1786 and one of Glenturret's rivals for the title of Scotland's oldest distillery.

BARONS AND BLENDERS

ON THE EVE of the passage of the Excise Act there were 111 licensed stills in Scotland, forty-two of them in the Highlands. Some of them – or at any rate their names, for the buildings are long gone – are still with us: Glenturret, Glen Garioch, Linkwood, Strathisla, Brackla, Clynelish, Tobermory, Bowmore, Lagavulin, plus a handful of others. Only three years later there were 263, of which 107 were in the Highlands; in the same period Britain's official spirits production nearly doubled to just over 18 million proof gallons or, in modern parlance, 80 million litres of pure alcohol.

For the existing legal distilleries, especially those in the Highlands, the lifting of trading restrictions that was part of the Excise Act was transformational: some that had previously limped along, such as Teaninich, now blossomed; others that had been forced to close, such as Blair Athol, now reopened. Many, if not most, of the 'new' distilleries weren't so new, however. To more enterprising smugglers, going legal was both attractive and easy. All you had to do was hand in your old pot and get a new one with a capacity of 40 gallons or more.

True, you had to pay both a licence fee and duty; but all the risk, interruption and inconvenience of the past were over. Some of the best-known distilleries of today have shady pasts: Cardhu was notorious as the home of Helen Cumming, who, when gaugers approached, would invite them in for tea and then make an excuse to go outside and wave a red flag to warn her husband, John, of their presence. Edradour, Scotland's smallest distillery, was founded legally in 1825 by a cooperative of local farmers to supply their own needs; but mysteriously, they seem to have been already experienced stillmen!

John Dewar, born to a poor crofting family in 1806, worked hard at school, showed promise, and got a job with a wine merchant in Perth at the age of twenty. Twenty years later he set up on his own and was among the first to bottle a proprietary blend – the legendary White Label – and to engage sales representatives in London.

Not everything was rosy for the Highlanders, however. Convictions for illegal distilling fell from 4,500 in 1823 to a mere 873 in 1825; but every household where distilling was given up lost an income estimated at 10s a week. There was real want, which occasionally flared into violence. Arson attacks on the new distilleries were frequent, and George Smith, founder of The Glenlivet, who had been an illegal distiller before the Act and was the descendant of generations

Opposite:
The Glenlivet Distillery, founded by George Smith.

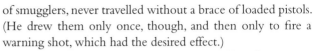

of smugglers, never travelled without a brace of loaded pistols. (He drew them only once, though, and then only to fire a warning shot, which had the desired effect.)

But although whisky was now respectable, elbowing its way into the mainstream drinks trade was not easy. As newcomers to an established market where the working classes already had their beer and gin and the aspiring and upper classes their port and brandy, the whisky distillers had to find wholesalers, retailers and a buying public for what was, to most, a novelty. They were helped by the fact that top-quality Scotch had already penetrated the uppermost strata of society

Poacher turned gamekeeper: George Smith of Glenlivet was one of the first of the 'smugglers' to go legitimate after the 1823 Excise Act.

– was it, one wonders, a well-placed Scottish peer who introduced the Prince Regent to The Glenlivet? – and soon after the Excise Act we find Scotch whisky on sale as a luxury item in the West End of London. In *Sketches By Boz* (1836) Charles Dickens, who used liquor of all sorts as a recurring motif throughout his works, had two City clerks on a gala night out in the Strand treating themselves to 'goes of the best Scotch ... and the very mildest cigars'. After that he fell silent on the subject of whisky, although his own cellar, according to his probate inventory, was well stocked with it. The best malts might have earned their place in the decanters of the wealthy, but in the mass market Scotch was slow to take off.

The first hurdle the distillers faced in bringing Scotch to market was the variability not only in its quality but in its very nature. Highland whisky was reputedly all malt; but while there may only rarely have been surpluses of other grains to pad out the wash, there are records of other fermentables, especially potatoes, being used when over-abundant. The malting process itself was often crude: sacks of barley were soaked in bog water, allowed to sprout and then parched as quickly as possible over wood or peat fires. There are also records of small-scale distillers continuing to use botanicals more generally associated with gin – especially herbs and wild flowers, but on occasion juniper berries and orange peel. Hurried distillation meant that some Highland products were high in heavy alcohols, especially fusel oil. And maturation was haphazard and little understood. The breadth of the spectrum, ranging from young, pale and aromatic to dark, heavy and fusel-rich, must have been quite enough to leave unfamiliar consumers utterly bemused.

The second hurdle was adulteration. Throughout history foodstuffs have been padded out with cheap stretchers, both legal and illegal. Victorian consumers seem to have taken it for granted that whatever they bought might contain more or less anything from the innocuous to the lethal; and whisky was no exception. Like so many commodities, it was shipped in casks,

A whisky dynasty – six generations of the Grant family, owners of the Glenfarclas distillery in Speyside since 1865 (left to right: John, George, George, George, John, and George).

which made it vulnerable to tampering. The agent might make three barrels out of two; the retailer might buy two of those barrels and make another three. An astonishing array of colouring and flavouring agents was conscripted to disguise the fact that the whisky had been so watered down. Cheap sherry, prune wine, green tea, glycerine, angelica root, cayenne pepper, various fruit concentrates – these were harmless. But methylated spirits, turpentine, even sulphuric acid, also common additives, were anything but. Governments from the mid-nineteenth century on legislated repeatedly against adulteration; but with whisky agents and retailers (rather than distillers, who were in the main not involved) they ran into a brick wall. There was no definition of whisky, the trade argued; traditionally it might carry high levels of fusel oil as well

as various additives and adulterants, or on the other hand it might not. Either way, there was no case to answer. A determined attempt to expose adulteration as a scandal came in 1872 when the *North British Daily Mail* hired a chemist to analyse more than thirty samples from Glasgow pubs. Many were entirely synthetic and contained no whisky whatever; of only a single sample could the chemist report 'a good whisky, free from adulteration save with water'. The investigation provoked much lively debate but, in the end, no action.

The third hurdle was commercial. The Lowlands possessed some large concerns making mostly grain spirit, much of it destined to become gin. Highland distilleries, though, were mainly small affairs whose operators had little or no experience of trading (legally) beyond their own backyards and no

Workers at
Glenfarclas,
c. 1895.

understanding of the markets now opening up to them. Nor can they have been technically well-equipped to meet the new challenges. Small manufacturers in any field rarely have the resources to analyse and improve their processes and to achieve the consistency that markets require. Highland

distillers by and large worked with the ingredients nature provided and the processes their parents passed down to them. They were not even in full control of the quality and consistency of their product when it left the distillery, still less as it made its way down the supply chain.

The answer to many of these problems lay in an Irish invention, the continuous or column still, which could produce a more or less endless stream of nearly pure alcohol and was to lead to the development of blended scotch. First patented in 1822 by Sir Anthony Perrier of the Spring Lane Distillery in Cork, it wasn't an immediate success but attracted the attention of Robert Stein of Kilbagie, a big Lowland distillery near Kincardine. Stein, an in-law of the powerful Haig dynasty, was already selling spirit on the English market. In 1826 he demonstrated an improved but still imperfect version of Perrier's still; watching was an Irish excise officer, Aeneas Coffey, who spotted how the device could be improved yet further and in 1830 took out his own patent and started manufacturing Coffey stills in Dublin. The Coffey still, highly efficient both mechanically and in terms of malt tax, was immediately adopted by distillers in both England and the Lowlands but was met with disdain by the Highland malt distillers as producing a spirit with no character. Yet it became, in the 1850s, the essential piece of technology in Scotch whisky's battle to reach an English and indeed an international mass market.

One of Scotch whisky's early problems, as we have seen, was that it ran such a broad gamut of expressions: even the same distillery's output could vary from year to year. Merchants such as Arthur Bell of Perth and Andrew Usher of Edinburgh therefore started 'vatting' or blending different vintages from the distilleries they represented to produce at least some degree of uniformity and, they hoped, an identity that the consumer could recognise. The availability of neutral spirit from the new continuous stills enabled the blenders to drown out or at least tone down jarring and overdominant notes – especially peatiness – to create a smoother drink that was both more consistent and more palatable to the Sassenach. In theory the admixture of 50 per cent or more of the much cheaper grain spirit would also bring the price of whisky within the reach of the common man; but this didn't happen until 1860, when the Spirit Act allowed maturing stocks to be held duty-free in bond. The cash-flow advantage of being allowed to defer payment of duty until the matured spirit went on sale was a game-changer for the big whisky merchants: it allowed them the leeway to mature their stocks properly and also led to a major innovation: bottling.

Opposite:
John Alexander Dewar, son of John Dewar, who ran the family business, eventually becoming first Baron Forteviot.

Bottling is always held out as the big blenders' answer to the problem of adulteration, and there is some truth in that. But it did more than give the blender control over quality: it created an affordable small package where previously the consumer had had to buy from the barrel; and it allowed the

creation of an eye-catching bottle label and a memorable brand name that the consumer could rely on. There's some argument as to which of the whisky merchants of the 1860s was the first to bottle and brand their own blends. Was it Usher's, Dewar's, John Walker, George Ballantine, Bell's or John Haig? They all claim the honour; they certainly all reaped the rewards.

For this was the beginning of a forty-year golden age, due not only to the efforts of the industry itself but also to external factors. Scotland's railways were connected to the English network in 1849; the lines that were built throughout the country in the following forty years brought coal and barley

to many a distillery siding, and carried casks of finished whisky away to the blenders. The Strathspey Railway, for example, opened in 1863 and linked distilleries at Forres, Cromdale, Advie, Ballindalloch, Knockando, Dailuaine, Aberlour and Craigellachie to the Keith–Dufftown line, which had opened the year before. The trading climate was improved still further in 1880 when Gladstone's Liberal government finally repealed the malt tax. The loss of revenue to the Exchequer was clawed back through higher duty; but again, cash-flow was greatly improved by deferral of payment.

Despite all these advances, the English remained stubbornly wedded to brandy until the advent of phylloxera.

Aberlour station on the Strathspey Railway, now the Speyside Way. The spreading of the rail network in the 1860s was crucial to the whisky industry's national and then global success in the 1870s and '80s.

This tiny aphid arrived in France from the United States in 1862 and fastened itself on to the country's vine roots, into which it injected lethal toxins. The unwelcome migrant spread rapidly and within a very few years had destroyed 40 to 50 per cent of France's vineyards. The cause of the blight was identified by 1868, and in the early 1870s the French began ripping out their old vines and replanting with American rootstock, which had a degree of immunity. But the vineyards of Cognac had been devastated and would take more than a decade to recover. By the mid-1870s brandy was in short supply. Scotch wasn't.

This opportunity coincided with the emergence of a new generation among the whisky dynasties, a generation whose businesses were on secure foundations and who were eager to find new markets. In the 1880s these confident young plutocrats established offices in London and set about making their presence felt. Thomas (always known as 'Tommy', even after he was ennobled) Dewar, younger son of John Dewar of Perth, was sent to London in 1885 aged only twenty-three. Refused permission to exhibit his White Label blend at a brewers' show at the Agricultural Hall, he fetched his bagpipes and paraded up and down outside playing as loudly as he could and refusing to stop, however earnestly he was begged to. It didn't get him into the show, but it did get him noticed; shortly afterwards Dewar's became sole supplier of Scotch to the high-society caterer and wholesaler Spiers & Pond.

Opposite:
In 1897 Dewar's built a distillery of its own, Aberfeldy in Perthshire, which still flourishes today.

The company was able to lease a distillery – Tulliemet in Perthshire – in 1890 and built one of its own – Aberfeldy, also in Perthshire – in 1896.

Then there was the dapper James Buchanan, who had been McKinlay's London agent before setting up on his own in 1884. He had the sense to choose an accountant who was also chairman of United Music Halls, all of whose bars were soon stocking his blend. He also made friends with the manager of a prestigious hotel, whom he cultivated for several months before dropping into the conversation that he was a whisky merchant. That account, too, was soon his. His final coup was to secure the account of the House of Commons. Buchanan's whisky was therefore given a sombre new bottle-dressing with the appropriate gravitas, from which it soon earned the nickname Black & White. (The image of terriers came later.)

Dewar and Buchanan were only two of the whisky paladins who cut a faintly and pleasingly exotic dash in London's social and business networks in the 1880s and '90s. Alexander Walker, son of the original Johnnie Walker (who, in contrast to the debonair image created by commercial artist Tom Brown in 1908, was rather a cautious and reserved man) was another; Peter Mackie of White Horse fame was a fourth. But capturing London was only the first step: the city was

Opposite:
A White Label
bottle, dating from
c. 1910.

James Buchanan
set up in business
in London in 1884.
He once booked
a table for twelve
at a hotel that he
knew didn't stock
his brand. When
the waiter arrived,
they all asked
for Buchanan's.
On hearing there
wasn't any,
they got up and
marched out.

also a gateway to the world. Not only was London the capital of that huge portion of the globe that was then coloured red to represent the British Empire on the maps; it was also the world's greatest entrepôt of finance and commerce and therefore a node from which whisky's reputation could spread. Success was not inevitable, however, and whisky barons had to work hard to cultivate world markets – in 1892 Tommy Dewar began a world tour that netted agents for White Label in twenty-six countries but took two whole years – but still, the world was ready for them. The Empire, as a sales territory, could be more or less taken for granted; Europe was still short of brandy; and the aspiring high society of North America regarded its native whiskies as too rustic for its taste and took eagerly to Scotch.

In 1886 William Grant, manager of Dufftown's Mortlach distillery, bought a complete set of second-hand equipment from Cardhu, which was re-equipping, to found Glenfiddich, which he ran with his wife, Elizabeth, and their three sons.

This expansion in sales was not only great; it was rapid. There was a risk of running out of the malt whiskies the blenders needed, and their response was to secure their supplies by launching what amounted to a reverse takeover of the malt distillers. In 1884 Leith-based blender William Sanderson bought one of Scotland's oldest distilleries, Glen Garioch, which had been established in 1797 and had remained in the hands of the founding family until that time. He re-equipped it with the two biggest pot stills in Scotland,

Workers at Glenfiddich in the late 1890s.

Previous page:
William Grant built
the Glenfiddich
distillery virtually
single-handed.

and from then on much of its output went into VAT 69. In 1891 Mackie's, in partnership with the Speyside whisky merchants Alexander Edward, built Craigellachie. In 1892 the Glasgow merchant Robertson & Baxter acquired one of its key suppliers, Glenglassaugh. In 1893 Johnnie Walker bought the famous Cardhu distillery. And so it went on.

The blenders were now undisputed masters of the industry — indeed, the sale of single malt almost died out.

They didn't just buy distilleries: they built them, too. In 1885–7 the journalist Alfred Barnard toured all 129 working Scottish distilleries; had he made the same tour in 1900 he would have had to visit thirty more, including such well-known names as Craigellachie (1891), Balvenie (1893), Glen Moray (1897) and Knockando (1899). Between 1886 and 1897 the small Speyside town of Dufftown acquired four of its seven distilleries. But that accounts only for brand-new

By 1890 Grant was doing well enough to build a second distillery, Balvenie, just down the road; today William Grant & Sons is the biggest family-owned company in the business and is a world-class competitor.

distilleries; in the same period dozens of existing ones were expanded or rebuilt. The extent of the investment in production can be gauged by the fact that the architect Charles Doig of Elgin (designer of the 'pagoda' maltings so emblematic of the Scottish distillery) was alone responsible for building or rebuilding a total of fifty-six distilleries.

Suspicions that this breakneck growth might be no more than a bubble seemed confirmed in 1898 with the collapse of the business owned by the brothers Robert and Walter Pattison, whisky merchants of Leith. Latecomers to the boom, the Pattisons fell into the hiatus between the satiation of actual demand and the moment that the banks realise the game is up. The two brothers were therefore able to borrow freely – squandering much of their capital, said disapproving contemporaries, on luxury and display – but could capture market share only by selling an inferior product cheaply and squeezing as much credit out of their suppliers as they could. When the banks finally called their loans in, a lot of malt distilleries were left holding bad debts and were forced to follow the brothers – both of whom were jailed for offences including paying dividends out of capital – into receivership. Between 1900 and 1908 sixty distilleries had to close, including some, such as Caperdonich, that had only just opened. Worse, the industry was holding more than 13 million proof gallons in bond, enough to satisfy a third of Britain's annual thirst for spirits.

But it wasn't a bubble. The supply side, especially the independent malt distillers, may have been fragmented and chaotic, as evidenced by the lawsuit taken out in 1880 by Glenlivet against the host of competitors who were using its name. (It ended in a compromise: as Glenlivet was the only distillery in the eponymous parish it remained entitled to call itself The Glenlivet, but its neighbours were still allowed to use the name in composites such as Dufftown-Glenlivet.) But the industry as a whole was solid: demand was there and the trade was profitable. In the preceding years a new breed of multiple conglomerates had been forming, and they were well-capitalised and businesslike, unlike the Pattisons. The independent sector had been due a shake-out, but the multiples were strong enough to survive.

The first and biggest of them was the Distillers Company Ltd, or DCL as it became known. It originated in 1856 when the six largest grain distillers formed a trade association – frankly, a cartel – to fix prices and carve up the market between them. The best-known, although not the largest, of DCL's founding companies was one of the Haig family businesses. In 1877 DCL was incorporated as a limited company, and it was floated on the London Stock Exchange in 1886. In response, the other grain distillers formed the rival North British Distillery Company, but DCL was always the stronger and, on the collapse of Pattison brothers, was able to buy their Leith warehouses, newly built at a cost of £60,000, for just £25,000. It had already built a malt distillery of its own, Knockdhu in Aberdeenshire, in 1894. During the contraction that followed the Pattison collapse DCL used its capital to diversify into grain distilling in England and Ireland and into the production of yeast and industrial alcohol.

The other big blenders also spent the first years of the twentieth century consolidating their positions, mainly through financial restructuring and the development of marketing and advertising campaigns. How far they had succeeded was about to be put to the sternest possible test.

The collapse of Pattison brothers, the First World War, Prohibition in the United States and finally the Depression resulted in the closure of all thirty distilleries in Campbeltown, including Benmore (shown here). Only three – Springbank, Caol Ila and Glengyle – ever reopened.

RECESSION AND RECOVERY

When war broke out in 1914, privations were only to be expected; but nobody can have imagined how great they would be. The Boer War a decade earlier had caused the industry scarcely a ripple. True, it all had to be paid for eventually, and Lloyd George's 'People's Budget' of 1909 had increased the duty on spirits by a third. It was just the industry's luck that Lloyd George – who conducted a long affair with his secretary, and was involved in a financial scandal, but who was evangelically teetotal – should still be Chancellor of the Exchequer when the First World War erupted. To their horror, they learnt that he was seriously entertaining complete prohibition, and, although they talked him out of it, they were forced to make concessions. Fortunately one of his chief advisers, James Stevenson, was one of their own, having been seconded to the Ministry of Munitions from Johnnie Walker. At a meeting in 1915 the whisky barons, led by DCL's formidable chairman, Harold Ross, learned of the forthcoming Immature Spirits Act, an idea of Stevenson's to preserve existing stocks by enforcing a minimum of three years' maturation in bond before sale. (This measure has since become one of the industry's quality benchmarks and is something it is very proud of.) They also agreed to convert six grain distilleries to acetone production and to supply industrial alcohol from the remainder. It seemed they had got off lightly.

But worse was to come as the war dragged on. Output was cut to half its 1916 level. The maximum permissible alcoholic strength was all but halved. In 1917 duty was increased to 30s per proof gallon – before the war it had been 14s 9d. The big grain distillers with their Government contracts, their stocks in bond and their comfortable balance sheets could scrape by, although Buchanan's and Dewar's merged in 1915 just to be on the safe side. But the independent malt distillers couldn't. One by one they closed, leaving the blenders anxious as to where their malt fillings were to come from. In 1916 DCL bought Speyburn and Tobermory while Dewar's took over Lochnagar and Talisker; but there were still concerns over sourcing fillings when the war was over.

They needn't have worried. Peace was followed by an economic boom, but demand for the industry's dwindling stocks (padded out by the discovery of a Government reserve of 3.5 million proof gallons) was held back by further increases in duty in 1919–20, which meant that a bottle of Scotch that cost 4s before the war now cost 12s 6d; and, of course, by Prohibition in the United States. And the boom proved short-lived: by the mid-1920s it had fizzled out almost everywhere

Hazelburn distillery in Campbeltown, now sadly closed.

except in the south-east of England, where modern industries such as electronics and aero-engineering were still blossoming. The Wall Street Crash of 1929 was followed by the Great Depression, and consumer demand was almost wiped out.

The 1920s and early '30s took a terrible toll on the independents. Campbeltown in Kintyre had had thirty-four distilleries over the years; at the end of Prohibition every one of them was closed, and only three have since reopened. The 133 working distilleries of 1920 fell to seventy-two by 1931 and fifteen by 1933. Of course, the majority of these silent stills were not dead, but sleeping: the comparative simplicity of distillery equipment meant they could be mothballed with

The sheriff's deputies dump illegal liquor during Prohibition, Santa Ana, California, 1932.

little maintenance; and their often remote locations, together with their size and specialised nature, made them hard to sell for other uses. Nonetheless, the independent sector was all but wiped out during the years 1925–33, with only the hardier companies such as William Grant of Glenfiddich fame, Highland Distillers, the Grants of Glenfarclas and a handful of others clinging on.

It was a different story for DCL. With its broad spread of interests and strong finances it was able both to catch the more desirable malt distilleries as they fell, acquiring Glendullan,

Glenlossie, Caol Ila, Clynelish, Old Pulteney, Balmenach and many others, and also to mop up all its major competitors. In the early 1920s it bought out the two independent companies still owned by the Haig family – Haig & Haig in 1922 and John Haig in 1924. In 1925 it snapped up both Scottish Malt Distillers (formed in 1914 through the merger of Glenkinchie, Clydesdale, Grange, Rosebank and St Magdalene) and even Buchanan-Dewar's. In 1926 it bought John Walker; in 1927 it bought Peter Mackie's White Horse Distillers.

DCL didn't quite reign supreme, however. For the first time, foreign money was coming into the stricken industry (unless you count English money as foreign: the gin distiller Gilbey's had bought Glen Spey, Strathmill and Knockando between 1887 and 1904). The money wasn't exactly clean, though: much of it was mob money.

A bootlegger's still, preserved at Havre Underground Museum, Havre, Montana.

Vast amounts of Scotch were smuggled into Prohibition America by blockade runners based in the Bahamas and St Pierre-Miquelon. But by the late 1920s the sea routes were becoming too dangerous, and the focus switched to Canada and the Detroit River narrows separating Windsor, Ontario, from Detroit. The man who sold the Detroit mobsters their stock was a legitimate tycoon, Samuel Bronfman, who in 1928 secured his source of supply by buying Seagram Distillers of Windsor, which had a stake in the Milton and Glenkeith distilleries and the Aberdeen merchant Chivas Brothers. (Chivas was already deeply involved in smuggling, even packing its bottles in watertight cases.) Bronfman sensed that Prohibition was ending and that whoever had footholds in both Scotland and the United States would profit handsomely. The same logic appealed to another big Windsor distiller, Hiram Walker, which bought the Glenburgie-Glenlivet distillery in 1930 and continued to invest throughout the 1930s, buying Ballantine's in 1936 and building the Inverleven grain distillery in 1938. These interventions were not hugely significant in themselves, but they prefigured the post-war era, when the Scotch whisky industry became a very tempting target for foreign investors.

The late 1930s were comparatively kind. The end of Prohibition and Roosevelt's New Deal created demand in the United States, while rearmament by all the Western powers generated jobs and disposable income at home. Many distilleries

Opposite:
New York detectives Moe Smith and Izzy Einstein celebrate the end of Prohibition with a (legal) bottle of White Horse.

New York mafia boss Frank Costello, whose company Alliance Distribution acquired Leith whisky merchant William Whiteley after Prohibition. Whiteley's blends The King's Ransom and House of Lords had been popular with American bootleggers as they were very rich and malty and could be watered down for extra profit.

were brought out of mothballs, and the number at work rose from 1933's fifteen to ninety-two at the outbreak of the Second World War. But the 1939 grain harvest was the last to be sold on the open market for many years. When rationing was introduced in January 1940, Scotch wasn't on the list. But it was all but impossible to come by, for the government closed all the grain distilleries and slashed the allocation of barley for malting: none of the 1943 harvest was allocated for distilling at all, and in 1944 there was only a single distillery working. Another sharp rise in duty stood in for rationing: when a bottle of Scotch at 13s 6d represented a fifth of the average weekly wage, not many would queue up to buy it. Instead, as much whisky as possible was exported for dollars: when the SS *Politician* of *Whisky Galore* fame sank off Eriskay in late 1941 she was carrying 28,000 cases of Scotch, or a third of a million bottles, all destined for the United States.

For Winston Churchill was no Lloyd George, and intended all along to preserve the industry. 'On no account reduce the barley for whisky,' he said at one point. 'This takes years to mature and is an invaluable export and dollar producer.' And the post-war Labour government agreed: in 1945 it increased the industry's grain allocation to 43 per cent of 1939 levels, but deterred home consumption with a series of swingeing duty increases that topped out at £10 10s 6d per proof gallon. In 1949 the grain allocation was restored to its 1936 level, but not until 1953 was wartime regulation abolished altogether.

It was the beginning of a new boom as the distilleries shook off their wartime lethargy and got back to work. The 1945 harvest supplied enough malt for fifty-six malt distilleries to resume production and by 1950 the number working had reached ninety-five – three more than in 1939.

And suddenly the Canadians were back, building on the strategic position that dated back to Prohibition. In 1950 Seagram's bought the Strathisla distillery in Keith. In 1954–5 Hiram Walker bought Glencadam in Brechin, Scapa in Orkney and Pulteney in Wick. In 1956 American investors followed the Canadians when Schenley Industries of New York bought an old-established English gin distiller, Seager Evans, which since 1927 had put together a Scottish mini-empire comprising the merchants Long John, the Glenugie malt distillery and the Strathclyde grain distillery. Schenley had deep pockets: in 1957 it opened a new malt distillery, Kinclaith, inside the Strathclyde complex, and in 1958 it built Tormore in Speyside. The following year another American-backed investor, Inver House, built both a grain and a malt distillery, Glenflagler, near Airdrie.

With prosperity returning to the West, Scotch was becoming the world's favourite spirit: more accessible than Cognac, more sophisticated than gin, more cosmopolitan than vodka, and with a range of qualities and prices suited to everyone from worker to jet-setter. By 1961 production had returned to pre-war levels and exports were increasing rapidly – £81 million in

1962, £228 million in 1972, £872 million in 1982. Home demand was boosted by the deregulation of off-licences in 1964, which spurred the growth of the supermarkets. To cope with this new demand, which in the whisky industry has to be estimated several years in advance, more than a dozen new distilleries were built in the 1960s and '70s, most of them to provide fillings, but some – Isle of Jura, Glenallachie, Tamnavulin – destined to become well-known single malts in their own right. This scale of building hadn't been seen since the 1890s, but in terms of output it was nothing compared to the extent to which existing distilleries were expanded. Balvenie had new stills added in 1957, 1965 and 1971. Glenfarclas went from two to four stills in 1960 and to six in 1976. Glengrant went from four stills to six in 1973 and to ten in 1977. And so it went on. Fettercairn, GlenDronach, Glen Garioch … the coppersmiths had never been so busy.

The expansion phase of the 1960s and '70s quickly attracted the attention of more outsiders. Some were foreign: French pastis giant Pernod-Ricard bought Aberlour in 1974; in 1980 Martini-Rossi acquired William Lawson Distillers, owner of Macduff, after a trading relationship that had already lasted thirty years; in 1981 Remy-Cointreau bought Glenturret. But at this stage most of the investment came from British brewers, who were well into the frenzy of amalgamation and takeover that created the Big Six. Not unnaturally, Scottish & Newcastle was first into the ring with

the purchase of Charles McKinlay in 1960. At the time McKinlay had already started rebuilding the old Isle of Jura distillery, which reopened in 1963; in 1968 it completed a new build at Glenallachie, intended to supply fillings for the McKinlay blend. In 1962 Courage bought the London wine merchants Saccone & Speed, which owned the Hankey Bannister blend. In 1972 Watney's bought International Distillers & Vintners, which included the three Scottish distilleries owned by Gilbey's. Three years later Whitbread bought Long John International, the trading name of Seager Evans. Last into the ring was Allied-Lyons, which laid the foundation for the enormous Allied Domecq global wines and spirits conglomerate when it bought William Teacher together with Ardmore and GlenDronach in 1976. (In fact the only Big Six brewer that didn't buy into distilling was the biggest, Bass. It had inadvertently found itself in possession of Auchentoshan when it bought the Glasgow brewer Tennent's in 1964 and sold the distillery on as quickly as it could.)

The boom couldn't last, and when the downturn of the early 1980s came it found the industry hugely overstocked and with a staggering surplus of capacity. Retrenchment became the order of the day. Between 1982 and 1985 twenty-nine distilleries – twenty-one of them owned by DCL – were either mothballed or closed for good. It was a sign of weakness that one corporate predator in particular was prepared to exploit. In 1985 Guinness bought the Perth distiller and blender Arthur

Opposite:
Tormore, Speyside, one of a crop of new distilleries that opened as prosperity returned in the 1960s. It was designed by Sir Albert Richardson, then President of the Royal Institute of British Architects.

Bell, whose main brand had overtaken the DCL-owned Haig as Britain's best-selling blend some years earlier. The following year it took over the mighty DCL itself, amid much controversy. Four of the participants, including the Guinness chief executive Ernest Saunders, were later jailed for false accounting; but the takeover stood, and Guinness became the world's leading spirits company. Allied, too, was on the hunt for a greater presence in the world's wines and spirits market: in 1984 it bought both United Rum Merchants and Hiram Walker, and in 1989 it acquired Whitbread's spirits business, including Laphroaig, the famously peaty Islay malt distillery.

The new regulations that followed the Monopolies & Mergers Commission's report into Britain's brewing industry in 1990 sparked a corporate meltdown in which the Big Six

Auchroisk at Mulben, Banffshire, was built in 1972–4 to supply malt fillings for IDV's J&B Rare. Designed to fit both its site and its heritage, its weatherproofing is traditional Scottish lime render and it incorporates small conical-roofed towers in the local architectural vernacular.

frantically reshuffled their assets to accommodate the new regime. These assets by now included well over half of the Scotch industry, which mostly fell into the hands of two supergroups: Pernod Ricard and Diageo. Allied merged with the Spanish wines and spirits giant Pedro Domecq in 1994 to create Allied Domecq, which was in turn bought by Pernod Ricard in 2005 for nearly £7.5 billion. Its whiskies are traded in the United Kingdom under the name Chivas Brothers. Guinness and Grand Metropolitan merged in 1997, with Guinness's wines and spirits subsidiary United Distillers and its Grand Met opposite number IDV together forming UDV. The conglomerate has now been renamed Diageo and owns, among many things, half the distilleries in Scotland, Britain's leading blend in Bell's, and the world's leading blend in Johnnie Walker.

Roseisle on the Moray Firsth, a vast state-of-the-art malt distillery opened in 2009 by Diageo to satisfy the thirst for whisky in emerging super-economies such as China, India and Brazil.

The creation of these two supergroups had two knock-on effects. The first was the appearance of other smaller but still formidable conglomerates, nearly all foreign-owned. Many of them were formed from distilleries that the two major players

The Famous Grouse with its younger siblings, The Black Grouse and The Snow Grouse.

had to discard to satisfy the competition authorities. They include Inver House (Pulteney, Balblair, Knockdhu, Speyburn, Balmenach), owned by a Thai company, International Beverage Holdings; Burn Stewart (Bunnahabhain, Deanston, Tobermory),

In 2001 Grant's revived the tradition of maturing its Founder's Reserve blend in old sherry casks. It subsequently started using ale casks as well.

William Grant & Sons was the first company to seek and win national distribution for a bottled single malt with the launch of Glenfiddich in the home market in 1961. An export version was launched in 1963.

Opposite:
Grant's Monkey Shoulder, launched in 2005, is what used to be known as a 'vatted malt'. It's a compound of Glenfiddich, Balvenie and Kininvie, blended without grain whisky.

bought by Distell of South Africa in 2013; Whyte & Mackay (Dalmore, Jura, Fettercairn), bought by United Breweries of India in 2007; John Dewar (Aberfeldy, Royal Brackla, Glendeveron), owned by Bacardi; Morrison Bowmore (Bowmore, Glen Garioch, Auchentoshan) bought by Suntory in 1994; and Glenmorangie (Glenmorangie, Ardbeg), part of the LVMH luxury goods group. Four other malt distilleries are individually owned by foreign investors, including Campari and Asahi; twenty-nine more are independently or family owned, the biggest concerns being William Grant (Glenfiddich, Balvenie) and the Edrington Group (Highland Park, The Macallan, Glenrothes, Glenturret, Tamdhu; also The Famous Grouse).

The second effect of the corporate chess-game was that in the 1970s whisky began to lose sales at home. The corporations had their eyes firmly on the bulk sellers: standard and premium blends. All of these were mature brands, in many cases approaching their centenaries; their quality and their tradition were rather taken for granted and there was little or nothing in the way of innovation to make Scotch appealing to new consumers. Much as traditional ales were losing ground to lager, dark spirits were being eclipsed among younger drinkers by the easy-drinking and very mixable vodka. Soaring exports – worth over £4 billion in 2012, with new markets such as China and Brazil catching up with North America and the European Union – were seen as more important than the home market. But the industry has shown much more willingness to innovate in recent years. In the 1980s Glenfiddich, hitherto almost the only single malt with nationwide distribution, was joined by others on the supermarket shelves, and The Glenlivet, The Macallan, Laphroaig, Cardhu and Highland Park became familiar names. In 1986 United Distillers launched its Classic Malt range. In the 1990s distillers started experimenting with different ages and expressions, rediscovering the method recommended by William Sanderson a century earlier of 'finishing' whisky in

sherry casks. More recently still, premium extensions of standard blends, such as Black Grouse and Snow Grouse, and rather less staid premium blends such as Monkey Shoulder have started appearing on the shelves.

And, even more encouraging, there have been a number of new distillery openings. Many of the newcomers – Loch Ewe, Daftmill, Speyside, Abhainn Dearg – are artisanal in inspiration and hark back to those distant times when distilling was a cottage industry, evoking in spirit, if perhaps not in their marketing material, the line from Robert Burns's poem 'Earnest Cry and Prayer': 'Freedom and whisky gang the gither!'

Opposite top: Kilchoman, the first new distillery on Islay since 1908. Founder Anthony Wills's aim is to source all his ingredients from the island itself.

Opposite bottom: Isle of Arran Distillers, opened in 1995, was one of the first of a new generation of privately owned distilleries aiming to recapture the values of artisanal rural distillers of the past.

Opened in 2008, Abhainn Dearg on the Isle of Lewis is the first licensed distillery in the Hebrides since 1840.

ENJOYING SCOTCH WHISKY

Whisky's versatility has long been underrated. Too many people think of it only as a short to be drunk with or without ice, with or without water, with or without soda or dry ginger. There's much more to it than that. Scotch is the basis of many classic cocktails; it has its role in cooking; and it certainly has a place at the dining table.

The quaich, a shallow silver bowl, is the traditional Scottish tasting vessel but has fallen out of favour because it doesn't allow the whisky to be properly nosed.

The first element in Scotch's versatility is the vast range of weights and degrees of pungency available, from light, fresh blends to ancient and venerable malts as dark as port. Even the most experienced professional taster can't possibly know them all, and most whisky drinkers will have established a core of a dozen or so reliable favourites that broadly cover the range of variables.

Then there's the glass. In one sense it doesn't matter what you drink your whisky out of so long as the hole is at the top and not the bottom. But like any other drink, Scotch has three principal sensory aspects: sight, smell and savour. You want a glass that will enhance all three. The quaich is the shallow dish, normally silver, in which whisky was traditionally sampled; but as it neither lets light through nor collects aromatic vapour it's not favoured by serious tasters today. The usual whisky tumbler, perfect for everyday use, isn't the serious taster's preferred option either, however fine the crystal. It enhances the visual aspect, certainly, but allows the fumes to escape too easily. An inward-curving sherry copita or even a brandy balloon cover all three aspects best.

Next, there's the water, either fluid or frozen. Plenty of people prefer their whisky

The copita, or straightforward sherry glass, is recommended by tasters because its inward-curving lip retains aromatic vapours.

A Manhattan.

An old-fashioned.

with a splash or have it well-watered – that is, about 50:50 – while others say that watering good whisky is sacrilege. And while there's no arguing with taste, the fact is that water dilutes. It has its most marked diluent effect on the whisky's alcohol content, reducing its bite and unmasking the maltier flavours, which is not necessarily a bad thing. Ice does much the same but also slows down the evaporation of volatile aromatics, accentuating the other flavours still further.

Moving a step beyond Scotch on the rocks or Scotch and water, there are as many whisky-based cocktails as there are mixologists. Many of the recipes are American and were formulated with Bourbon in mind, so versions made with Scotch will be less sweet: a full-bodied blend such as Whyte & Mackay Special makes a good substitute for Bourbon. Many recipes also require a syrup of two parts sugar to one part water, boiled until the sugar dissolves and taken off the heat before it can thicken. An old-fashioned, one of the earliest recorded cocktails, was originally whisky and triple sec with Angostura bitters; for a more modern version cut a big piece of orange zest, squeeze it lightly and rub it vigorously over the bottom of a tumbler. Add a teaspoon of sugar syrup and a couple of drops of Angostura bitters, then plenty of ice and finally a large Scotch. Serve with the piece of zest and a maraschino cherry. A whisky sour is one part sugar syrup to two parts whisky and two parts freshly squeezed lemon juice; with soda it becomes a Collins.

A Manhattan is simplicity itself: four parts whisky, one part sweet vermouth and a drop of Angostura bitters. There are dozens more, but these are perhaps the best known: better known still are the whisky toddy and Atholl Brose. Toddy is simply a shot of whisky with a teaspoon of sugar and a squeeze of lemon juice all dissolved in hot (not boiling) water. Honey is even better than sugar; rubbing the inside of the glass with the cut end of a ginger root first makes it better still. To make Atholl Brose, soak half a cup of oatmeal in half a pint of cold water; let it stand for an hour or two; strain it and feed the oatmeal to the chickens; mix four tablespoons of honey into the liquid; bottle it with as much whisky as you desire; stopper the bottle very tightly; and then shake it until the honey dissolves.

In cooking, whisky will substitute for brandy in any recipe such as cream, white wine and brandy sauce, or Chantilly. It's better than brandy in rich fruit-based cakes and desserts, including Christmas pudding and home-made mincemeat, and is essential in Dundee cake and, of course, home-made marmalade. And then

A whisky sour.

there's cranachan. Toast 3 ounces of pinhead oatmeal in a very hot frying pan, shaking a little from time to time to stop it sticking. As soon as it takes colour, pour it into a bowl or jar, cool it for a few moments and stir in a dessertspoon of whisky. Leave it for a few hours. Then whip up half a pint of double cream with another dessertspoon (or less) of whisky and fold the oatmeal into it. Into as many tumblers as you have guests put a few raspberries or – less traditional but considerably cheaper – strawberries, and spoon in the oatmeal, cream and whisky mixture. A light dusting of freshly toasted oatmeal (without whisky) and a few more berries tops it off nicely.

Strawberry cranachan.

Finally, whisky tops and tails a good dinner absolutely perfectly. As an aperitif, a fairly light blend or single malt on the rocks, especially a slightly peaty blend such as White Horse, can't be beaten. As a digestif, a rich malt – Bunnahabhain, say, or Highland Park – is as good as any Cognac and probably more memorable.

PLACES TO VISIT

Increasingly since the 1980s, Scotch and tourism have gone hand in hand. Visitors often take in a distillery as part of their exploration of Scotland's magnificent scenery and heritage; or it may be that looking at the scenery and learning about its heritage is a way of filling time between distillery visits. Either way, no Scottish holiday is complete without at least one distillery tour.

The Strathspey Railway, opened in 1863, is now a footpath for hikers linking some of Scotland's best-known malt distilleries through gloriously unspoilt countryside.

About half of Scotland's one hundred or so malt-whisky distilleries now welcome the public. Some have fully fledged visitor centres with cafés, gift shops and tours of different durations and prices; others offer only basic tours by arrangement. Some are open year-round; others only during peak holiday times. So, when planning your visit, always be sure to check the website and/or ring ahead.

LOWLANDS

Auchentoshan Distillery, by Dalmuir, Clydebank, Glasgow G81 4SJ.
   Telephone: 01389 878561. Website:
   www.auchentoshan.com
Bladnoch Distillery, Wigtown DG8 9AB.
   Telephone: 01988 402605. Website: www.bladnoch.co.uk
Glengoyne Distillery, Dumgoyne, near Killearn, Glasgow G63 9LB.
   Telephone: 01360 550254. Website: www.glengoyne.com
Glenkinchie Distillery, Pencaitland, Tranent, East Lothian EH34 5ET.
   Telephone: 01875 342012.
   Website: www.discovering-distilleries.com

HIGHLANDS (SOUTH)

Aberfeldy Distillery, Aberfeldy, Perthshire PH15 2EB.
   Telephone: 01887 822010. Website: www.dewars.com
Blair Athol Distillery, Perth Road, Pitlochry, Perthshire PH16 5LY.
   Telephone: 01796 482003.
   Website: www.discovering-distilleries.com

Dalwhinnie Distillery, Dalwhinnie, Inverness-shire PH19 1AA.
Telephone: 01540 672219.
Website: www.discovering-distilleries.com
Deanston Distillery, near Doune, Perthshire FK16 6AG.
Telephone: 01786 843010. Website: www.deanstonmalt.com
Edradour Distillery, Pitlochry, Perthshire PH16 5JP.
Telephone: 01796 472095. Website: www.edradour.com
Glenturret Distillery, The Hosh, Crieff, Perthshire PH7 4HA.
Telephone: 01764 656565. Website: www.thefamousgrouse.com

Glenturret at Crieff, Perthshire, claims to be Scotland's oldest distillery and, as well as offering tours ranging from standard to very de luxe, it has both a whisky school and a cookery school.

Tullibardine Distillery, Stirling Street, Blackford, Perthshire PH4 1QG.
Telephone: 01764 661800. Website: www.tullibardine.com

HIGHLANDS (EAST)

Fettercairn Distillery, Fettercairn, Laurencekirk, Kincardineshire
AB30 1YE.
Telephone: 01561 340205. Website: www.whyteandmackay.com

GlenDronach Distillery, Forgue by Huntly, Aberdeenshire AB54 6DB.
Telephone: 01466 730202.
Website: www.glendronachdistillery.com

Glen Garioch Distillery, Distillery Road, Oldmeldrum, Aberdeenshire
AB51 0ES.
Telephone: 01651 873450. Website: www.glengarioch.com

Glenglassaugh Distillery, Portsoy, Aberdeenshire AB45 2SQ.
Telephone: 01313 355131. Website: www.glenglassaugh.com

Royal Lochnagar Distillery, Balmoral, Crathie, Ballater, Aberdeenshire
AB35 5TB.
Telephone: 01339 742700.
Website: www.discovering-distilleries.com

HIGHLANDS (NORTH)

Balblair Distillery, Edderton, Tain, Ross-shire IV19 1LB.
Telephone: 01862 821273.

Clynelish Distillery, Brora, Sutherland KW9 6LR.
Telephone: 01408 623000.
Website: www.discovering-distilleries.com

Dalmore Distillery, Alness, Ross-shire IV17 0UT.
Telephone: 01349 882362. Website: www.thedalmore.com
Glen Ord Distillery, Muir of Ord, Ross-shire IV6 7UJ.
Telephone: 01463 872004.
Website: www.discovering-distilleries.com
Glenmorangie Distillery, Tain, Ross-shire IV19 1PZ.
Telephone: 01862 892477. Website: www.glenmorangie.com
Old Pulteney Distillery, Huddart Street, Wick, Caithness KW1 5BA.
Telephone: 01955 602371. Website: www.oldpulteney.com
Tomatin Distillery, Tomatin, Inverness-shire IV13 7YT.
Telephone: 01463 248144. Website: www.tomatin.com

HIGHLANDS (WEST)

Ben Nevis Distillery, Lochy Bridge, Fort William PH33 6TJ.
Telephone: 01397 700200. Website: www.bennevisdistillery.com
Oban Distillery, Stafford Street, Oban, Argyll PA34 5NH.
Telephone: 01631 572004.
Website: www.discovering-distilleries.com

ISLANDS

Abhainn Dearg Distillery, Carnish, Isle of Lewis HS2 9EX.
Telephone: 01851 672429. Website: www.abhainndearg.co.uk
Highland Park Distillery, Holm Road, Kirkwall, Orkney KW15 1SU.
Telephone: 01856 874619. Website: www.highlandpark.co.uk
Isle of Arran Distillery, Lochranza, Isle of Arran KA27 8HJ.
Telephone: 01770 830264. Website: www.arranwhisky.com

Jura Distillery, Craighouse, Isle of Jura PA60 7XT.
  Telephone: 01496 820385. Website: www.jurawhisky.com
Talisker Distillery, Carbost, Isle of Skye IV47 8SR.
  Telephone: 01478 614308.
  Website: www.discovering-distilleries.com
Tobermory Distillery, Ledaig, Tobermory, Isle of Mull PA75 6NR.
  Telephone: 01688 302647. Website: www.tobermorymalt.com

SPEYSIDE

Aberlour Distillery, High Street, Aberlour, Banffshire AB38 9PJ.
  Telephone: 01340 881249. Website: www.maltwhiskydistilleries.com
Benromach Distillery Co. Ltd, Invererne Road, Forres, Moray IV36 3EB.
  Telephone: 01309 675968. Website: www.benromach.com
Cardhu Distillery, Knockando, Aberlour, Banffshire AB38 7RY.
  Telephone: 01479 874635.
  Website: www.discovering-distilleries.com
Cragganmore Distillery, Ballindalloch, Banffshire AB37 9AB.
  Telephone: 01479 874715.
  Website: www.discovering-distilleries.com
Glen Grant Distillery, Elgin Road, Rothes, Aberlour, Banffshire
  AB38 7BS.
  Telephone: 01340 832118. Website: www.glengrant.com
Glen Moray Distillery, Bruceland Road, Elgin, Moray IV30 1YE.
  Telephone: 01343 550900. Website: www.glenmoray.com
Glenfarclas Distillery, Ballindalloch, Banffshire AB37 9BD.
  Telephone: 01807 500345. Website: www.glenfarclas.co.uk

Glenfiddich Distillery, Dufftown, Banffshire AB55 4DH.

Telephone: 01340 820373. Website: www.glenfiddich.com

The Glenlivet Distillery, Ballindalloch, Banffshire AB37 9DB.

Telephone: 01340 821720. Website: www.theglenlivet.com

The Macallan Distilleries Ltd, Easter Elchies, Craigellachie, Banffshire AB38 9RX.

Telephone: 01340 872280. Website: www.themacallan.com

Strathisla Distillery, Seafield Avenue, Keith AB55 5BS.

Telephone: 01542 783044.

Website: www.maltwhiskydistilleries.com

Dallas Dhu at Forres, Moray, was built in 1899 just as the whisky boom was coming to its near-catastrophic end. It had rather a chequered career, with recurring water-supply problems, and in 1983 was chosen by the Distillers Company for closure as part of the great production shake-out. However, it was rescued by Historic Scotland before it could be dismantled and is now a fascinating museum, offering a unique view of a classic Victorian distillery.

ISLAY

Ardbeg Distillery, Port Ellen, Isle of Islay PA42 7EA.
   Telephone: 01496 302244. Website: www.ardbeg.com
Bowmore Distillery, School Street, Bowmore, Isle of Islay PA43 7JS.
   Telephone: 01496 810441. Website: www.bowmore.com
Bruichladdich Distillery, Bruichladdich, Isle of Islay PA49 7UN.
   Telephone: 01496 850190. Website: www.bruichladdich.com
Bunnahabhain Distillery, Port Askaig, Isle of Islay PA46 7RP.
   Telephone: 01496 840646. Website: www.bunnahabhain.com
Caol Ila Distillery, Port Askaig, Isle of Islay PA46 7RL.
   Telephone: 01496 302769.
   Website: www.discovering-distilleries.com
Kilchoman Distillery, Rockside Farm, Bruichladdich, Isle of Islay PA49
   7UT.
   Telephone: 01496 850011. Website: www.kilchomandistillery.com
Lagavulin Distillery, Port Ellen, Isle of Islay PA42 7DZ.
   Telephone: 01496 302749.
   Website: www.discovering-distilleries.com
Laphroaig Distillery, Port Ellen, Isle of Islay PA42 7DU.
   Telephone: 01496 302418. Website: www.laphroaig.com

CAMPBELTOWN

Springbank Distillery, Well Close, Campbeltown, Argyll PA28 6ET.
   Telephone: 01586 551710.
   Website: www.springbankwhisky.com

FURTHER READING

al-Hassan, Ahmad Yousef. *History of Science and Technology in Islam*. Publication pending.

Barnard, Alfred. *The Whisky Distilleries of the United Kingdom*. Mainstream Publishing, 1987.

Bruce Lockhart, Sir Robert. Scotch: *The Whisky of Scotland in Fact and Story*. Putnam, 1951.

Burns, Edward. *Bad Whisky*. Balvag Books, 1995.

Buxton, Ian, and Hughes, Paul S. *The Science and Commerce of Whisky*. The Royal Society of Chemistry, 2013.

Darwen, James. *The Illustrated History of Whisky*. Harold Starke, 1993.

Forbes, R. J. *A Short History of the Art of Distillation from the Beginnings Up to the Death of Cellier Blumenthal*. E. J. Brill, 1948.

Gunn, Neil M. *Whisky and Scotland*. Routledge & Sons, 1935.

McDowall, R. J. S. *The Whiskies of Scotland*. John Murray, 1967.

Murray, Jim. *Jim Murray's Complete Book of Whisky*. Carlton Books, 1997.

Rosie, George. *Curious Scotland: Tales from a Hidden History*. Granta Books, 2004.

Smith, Gavin D. *The Secret Still: Scotland's Clandestine Whisky Makers*. Birlinn Ltd, 2002.

INDEX